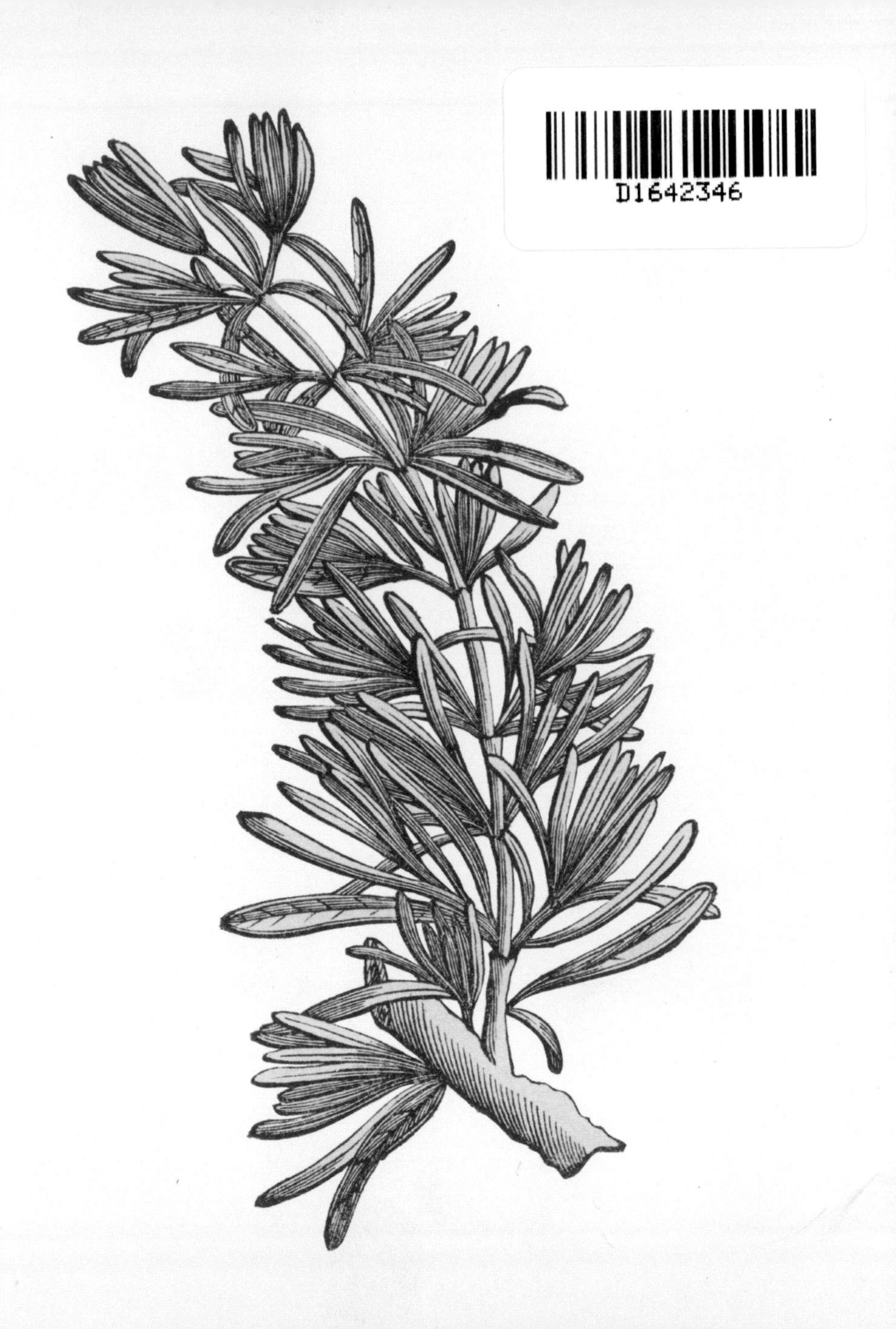
D1642346

KEW POCKETBOOKS

HERBS AND SPICES

Curated by Mark Nesbitt and Gina Fullerlove

Kew Publishing
Royal Botanic Gardens, Kew

THE LIBRARY AND ARCHIVES AT KEW

KEW HOLDS ONE OF THE LARGEST COLLECTIONS of botanical literature, art and archive material in the world. The library compri ses 185,000 monographs and rare books, around 150,000 pamphlets, 5,000 serial titles and 25,000 maps. The Archives contain vast collections relating to Kew's long history as a global centre of plant information and a nationally important botanic garden including 7 million letters, lists, field notebooks, diaries and manuscript pages.

The Illustrations Collection comprises 200,000 watercolours, oils, prints and drawings, assembled over the last 200 years, forming an exceptional visual record of plants and fungi. Works include those of the great masters of botanical illustration such as Ehret, Redouté and the Bauer brothers, Thomas Duncanson, George Bond and Walter Hood Fitch. Our special collections include historic and contemporary originals prepared for *Curtis's Botanical Magazine*, the work of Margaret Meen, Thomas Baines, Margaret Mee, Joseph Hooker's Indian sketches, Edouard Morren's bromeliad paintings, 'Company School' works commissioned from Indian artists by Roxburgh, Wallich, Royle and others, and the Marianne North Collection, housed in the gallery named after her in Kew Gardens.

INTRODUCTION

HERBS AND SPICES ARE REMARKABLE IN THAT they contribute little to the core essentials of nutrition, such as energy and vitamins, and yet we can hardly imagine food without them. Whether used alone or in combination, they bring complex tastes and aromas to sweet and savoury dishes and to beverages. Furthermore, some bring strong associations, such as the scents and flavours of festive or holiday foods. Throw in the medicinal benefits of many, and it is no surprise that there is a huge literature on herbs and spices; we highlight some further reading on pages 93–94.

What brings herbs and spices together is that both contain essential oils and other compounds, in sufficient quantity to be useful to humans. What divides them is that herbs are usually used fresh, in the form of leaves and occasionally stems, while spices are used dry (and sometimes fresh), and come from other parts of the plant, including the fruits and seeds, roots, barks and so on. Many spices come from the tropics, and it is therefore essential that they retain their aroma over long journeys, after drying. In contrast, herbs mostly lose their aroma on drying, with notable exceptions such as lavender and rosemary. Until the advent of airfreight, it was the gardeners' art that brought herbs close enough so that they could be used fresh. Herbs still lend themselves well to cultivation outside the kitchen door or on a windowsill.

Essential oils are complex mixtures of chemicals that share the characteristic of being volatile; in other words, they evaporate at room temperature. It is this volatility – the escape of molecules from the plant – that allows us to smell their fragrance. As well as being consumed in the form of herbs and spices, essential oils can also be extracted by distillation. These extracted oils are used to scent the home, and on a large scale in industry for cosmetics, food flavourings and medicine.

Botanists are still working out the benefits of essential oils to the plants that produce them, and it is clear that these vary from species to species. In many cases they are poisonous or repellent to predators, including microbes. Indeed, the anti-microbial effects of essential oils underpin many of their medicinal benefits to humans. At the same time, the poisonous properties are a reminder that many essential oils are toxic to humans too. In other cases, the aroma attracts pollinators and seed dispersers – it is no surprise that some herbs, such as lemon balm, are so attractive to bees. It is also speculated that essential oils play an antioxidant role in the plant, mopping up excess free radicals.

Terpenes are the most abundant components of essential oils and are the basis of such recognisable aromas as citrus, mint and conifer. Most essential oils are made up of a mixture of terpenes and other chemicals, and can vary significantly in their aroma and other properties. The composition is affected by the environment – for example, hot climates may stimulate greater production of these chemicals – and by genetics, leading to the breeding of varieties with specific properties. The gardener is well advised to give careful attention to which variety of a herb or spice they grow.

Easily portable and highly valuable, spices are well-adapted to long-distance trade. Both China and Europe have traded in spices for well over two millennia, with evidence for cinnamon (from India or points further east) in Greece by the 7th century BCE, and cloves from the Malaku Islands (Moluccas) in China by the 3rd century BCE. Following the Roman conquest of Egypt 2,000 years ago, there was both sufficient wealth and a secure trade route to support the import to ancient Rome of massive quantities of black pepper (page 83), as well as cinnamon (page 55), ginger (pages 4 and 67) and cloves (page 63). These crossed the Indian Ocean, arriving at the Red Sea and travelling overland to the Mediterranean. At about the same time, the overland Silk Road developed through central Asia and Iran.

From about 1100 the spice trade with Europe was revived through Indian and Arab seaborne trade, enriching Constantinople and the north Italian cities of Venice, Genoa and Florence. During the 15th century European powers began looking for direct sea passage for their own ships to the spice lands of India and South-East Asia. Portuguese traders used trading posts on the coast of West Africa, leading to a period in which African spices such as Ethiopian pepper (page 12) were much used in Europe. Following Vasco de Gama's circumnavigation of the Cape of Good Hope and arrival in India in 1498, the Portuguese took control of territories in India and Indonesia, establishing partial control of the spice trade. In 1492 Columbus set sail westwards, in search of a direct route to the Indies, but stumbling upon the Americas instead. The Portuguese were displaced in Asia by the Dutch, and then the British East India Company. The romance of spices, dating to an era when their origins were unknown, was replaced by a reality of plantation agriculture, slavery, and colonial government.

Today's international trade in spices is estimated to be worth over three billion pounds, with China and India as the leading exporters. The demand for spices continues to grow with the increased interest by consumers in international cuisine, and in the medicinal benefits of herbs and spices as part of a diverse diet. As a high value product, production of herbs and spices has the potential to make good returns for producers, who are often smallholders. Here the readers of this book can play a part, as consumers. If inspired to seek out new tastes, we encourage them to seek out Fairtrade or organic products, and to look to manufacturers for a commitment to ethical and sustainable production.

Mark Nesbitt
Curator, Economic Botany Collection
Royal Botanic Gardens, Kew

Nº 1

Xylopia aethiopica

Ethiopian pepper, grains of Selim
from Matthias de L'Obel
Plantarum seu Stirpium Icones, 1581

The Ethiopian pepper grows wild in the evergreen rain forests of Africa and is an important crop in Ghana. The pungent and aromatic fruits are dried, sometimes by smoking over a fire, and used as a spice and medicine. In West Africa they are typically added whole to soups, stews, sauces and porridge.

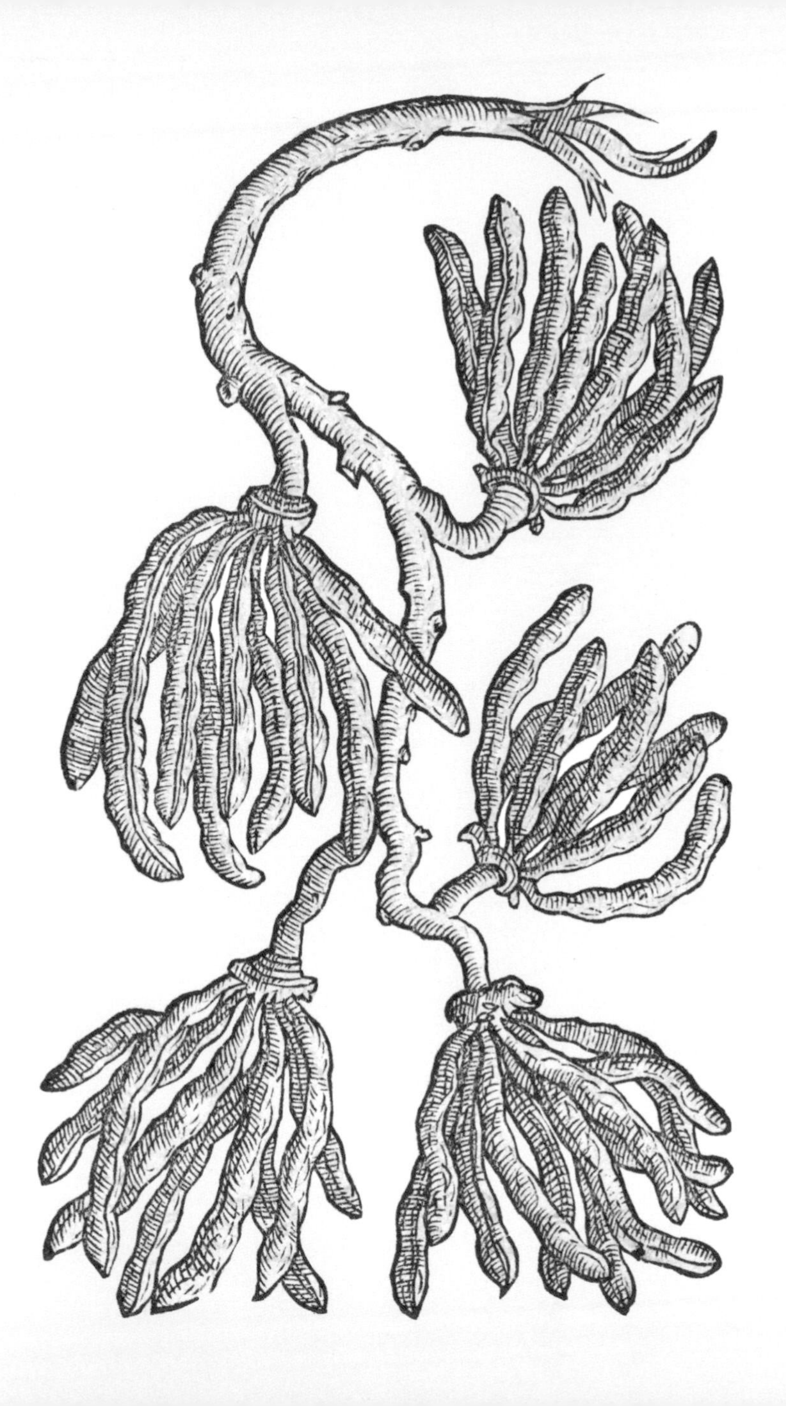

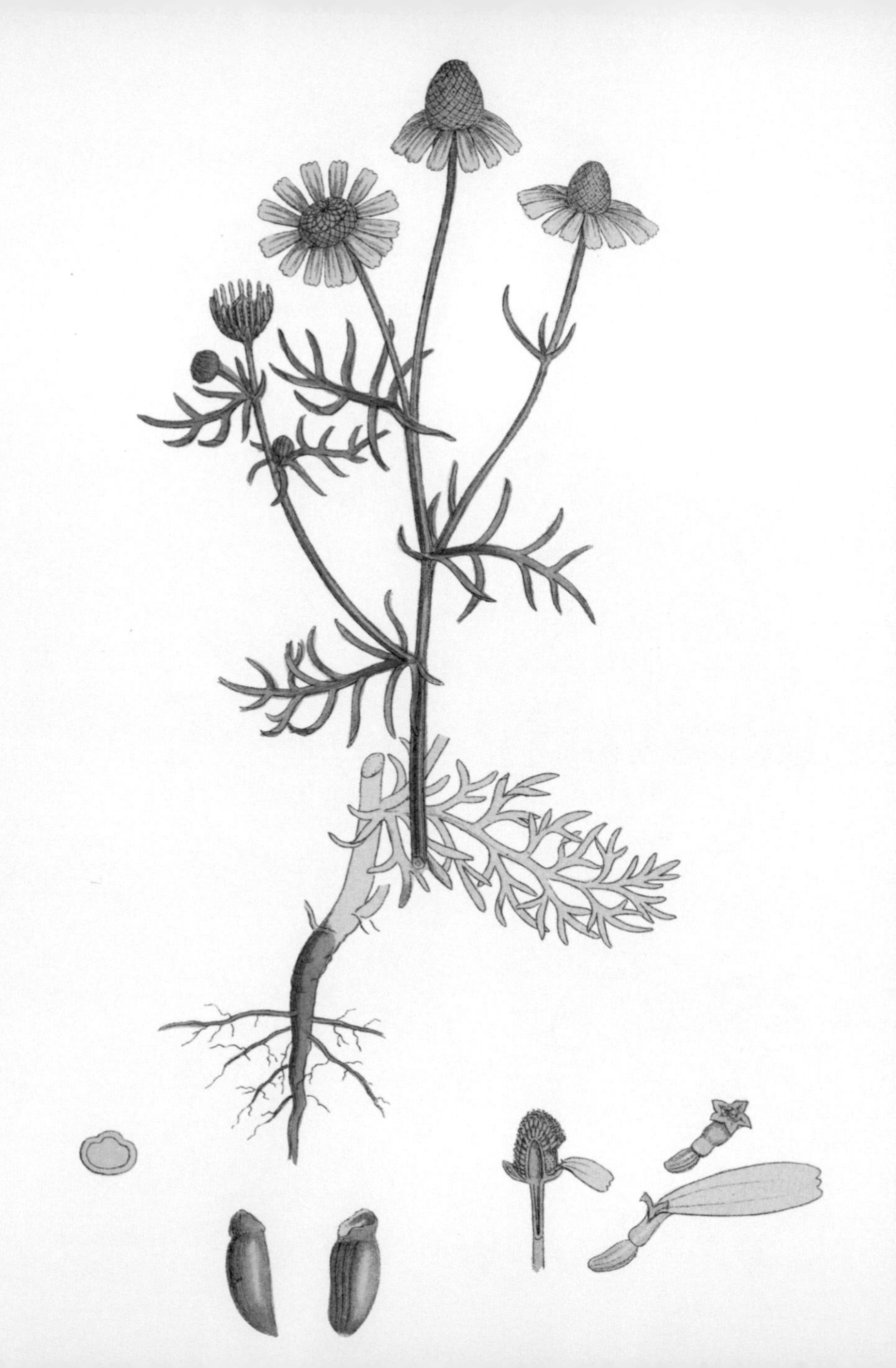

Nº 2

Matricaria chamomilla

German chamomile

from James Sowerby *English Botany*, 1866

This is one of two species of chamomile – the other being Roman chamomile (*Chamaemelum nobile*) – that are native to Europe and Asia; both are widely cultivated for their antimicrobial and digestive properties and as a relaxing tea.

Nº 3

Monarda didyma

bergamot, bee balm

from Joseph Plenck *Icones Plantarum Medicinalium,* 1788–1812

Bergamot is native to moist habitats in eastern North America and is cultivated in Europe as a fragrant garden plant. It has a long history of medicinal use by Native Americans, leading to the use by settlers of the young leaves as a tea. Bergamot should not be confused with the bergamot orange, the source of the essential oil that flavours Earl Grey tea.

Nº 4

Coriandrum sativum

coriander, cilantro

by Beari Lall commissioned by Adam Freer, Kew Collection, 1809

Coriander is both a spice and herb. The sweet, aromatic flavoured dried fruits are a key spice component in Indian curries and are important in European cuisine. The leaves, which lose their tangy aroma when dried, are used fresh with seafood and other dishes in South America and South-East Asia. Coriander was probably first cultivated in the Mediterranean region, at least 5,000 years ago.

Nº 5

Illicium verum

star anise

from François Pierre Chaumeton

Flore Médicale Décrite, 1815–20

The dried, star-shaped dark-brown fruits come from a slender evergreen tree, mainly grown in China and Vietnam. The warm, sweet scent is from the essential oil anethole, which is also the main flavouring in anise and fennel. The fruits are used in Chinese dishes of poultry and pork, often in a 'five spice' powder, while the essential oil is widely exported and used as a flavouring and aroma in industry.

Nº 6

Ocimum basilicum

basil

by unknown Indian artist commissioned by
William Roxburgh, Kew Collection c.1800

Basil was probably first taken into cultivation in the Mediterranean region and remains most popular in its cuisine. It is used fresh in many dishes including pesto, and as an accompaniment to tomatoes. Other species of basil are much used in South-East Asian cooking.

Nº 7

Zanthoxylum armatum

Sichuan pepper, Nepal pepper
by Matilda Smith from
Curtis's Botanical Magazine, 1918

The fruits of many species of *Zanthoxylum* are sold as Sichuan pepper; here we illustrate the Nepal pepper, *Z. armatum*, which grows wild throughout much of Asia. It is a common ingredient in Nepalese curries and pickles. Sichuan peppers contain sanshool, a chemical that elicits a pleasant tingling, numbing sensation.

Nº 8

Foeniculum vulgare

fennel

by unknown Indian artist commissioned by William Roxburgh, Kew Collection c.1800

Fennel is a multipurpose plant, with the swollen stem bases and leaves of Florence fennel used as a vegetable, and the seeds as a spice. Bitter fennel was first taken into cultivation in southern Europe, perhaps in classical times, and its seeds are used in eastern European cooking. Sweet fennel originated later and is the most widely used variety, with anise-scented, sweet and aromatic seeds.

No 9

Lavandula dentata

fringed lavender, French lavender
from *Curtis's Botanical Magazine*, 1798

This is one of the less common lavender species, with a native distribution in Arabia and the Mediterranean. It is grown in gardens and has a strongly aromatic scent. The leaves and flowers of *L. angustifolia* are added to salads and desserts, and the flowers are crystallised for use as cake decorations.

Nº 10

Melissa officinalis

lemon balm

by Elizabeth Blackwell from Elizabeth Blackwell *Herbarium Blackwellianum,* 1750–73

The leaves have a fresh, citrusy scent. Their main traditional use is medicinal, but culinary uses in Europe include fish dishes, desserts and as a garnish for drinks. The fresh or dried leaves are used as a tea. The essential oils are exceptionally volatile and easily lost on drying.

Nº 11

Borago officinalis

borage, starflower

from Joseph Plenck

Icones Plantarum Medicinalium, 1788–1812

The leaves have a mild cucumber flavour, sharing some of the same chemistry with the salad vegetable. The flowers can be used as a garnish for drinks and salads.

Nº 12

Rhus coriaria

sumac

by Elizabeth Blackwell from Elizabeth Blackwell *Herbarium Blackwellianum*, 1750–73

The fruits of sumac have a tart and astringent flavour owing to their content of tannins and organic acids. The dark-red dried and ground sumac fruit is sprinkled on many dishes in Iran and Turkey and is an ingredient of zaatar, a spice mixture used throughout the Levant. Sumac was used in classical Rome, before the lemon supplanted it as a sour flavour.

Nº 13

Petroselinum crispum

parsley

from F. E. Köhler *Medizinal-Pflanzen*, 1887

Parsley was taken into cultivation in the Mediterranean region, and it remains most popular as a culinary herb in Europe and the Middle East. Curly-leaved varieties are favoured in the British Isles for their decorative properties where parsley is used as a garnish; elsewhere, it is the flat-leaved forms that are chopped into dishes. The leaves retain little aroma after drying.

No 14

Mentha spicata

mint

from F. G. Hayne *Getreue Darstellung und Beschreibung der in der Arzneykunde Gebräuchlichen Gewächse,* 1805–46

One of the most commonly used of the many mint species, native to the Mediterranean and now widely cultivated. It is used in British cuisine for sauces, beverages and as a garnish. The refreshing aroma derives from essential oils including menthol.

N° 15

Capsicum annuum Longum group

chilli, paprika

by unknown Indian artist for Adam Freer, Kew Collection, c.1810

Five species of chilli pepper were taken into cultivation in Mexico and South America at least 6,000 years ago. Here we illustrate a variety of the most widely available species which is used both as a vegetable and spice. The hot, pungent flavour of *Capsicum* fruits derives from capsaicin. Chilli peppers are now used in many cuisines, following their rapid spread to Europe and Asia in the 16th century.

Nº 16

Salvia officinalis

sage
from Pierre Bulliard
Flora Parisiensis, 1776

The leaves of sage were first used in the Mediterranean region as a medicine, with their use as a culinary herb developing by the 16th century. The distinctive aromatic taste is robust and suits hearty dishes.

Nº 17

Angelica archangelica

angelica
from Joseph Plenck
Icones Plantarum Medicinalium, 1788–1812

The candied stalks of angelica are used in cakes and confectionary. Native to northern Europe, it is now grown more widely in Europe, especially France, for its stems, and also its roots which are used in herbal medicine and alcoholic beverages.

Nº 18

Thymus vulgaris

thyme

from H. L. Duhamel du Monceau

Traité des Arbres et Arbustes que l'on Cultive en France en Pleine Terre, 1800–19

This is the most widely used of several species of thyme. It originates in southern Europe but is now more widely grown. The strongly aromatic leaves are used in savoury dishes in Europe and North America, with key roles in the creole cooking of New Orleans, and the bouquet garni of French cuisine.

No 19

Curcuma longa

turmeric

by unknown Indian artist commissioned by William Roxburgh, Kew Collection c.1790

Turmeric was probably first grown in India and is now widely cultivated in tropical Asia and parts of South America. The plant belongs to the ginger family and as with ginger, it is the rhizome (root) that is consumed. It is valued for its pungent fragrance and for the yellow colour it gives to dishes. It is much used in Indian dishes and is reputed to have many health benefits.

Nº 20

Pimenta dioica

allspice
from Michel Étienne Descourtilz
Flore Pittoresque et Médicale des Antilles,
1821–9

The unripe fruits of allspice are gathered from wild and cultivated trees in its area of origin, in Central America and the Caribbean. The fruits are rich in the essential oil eugenol, giving a powerfully aromatic scent with echoes of clove (in the same plant family), pepper and other spices. In Jamaica, the largest producer, allspice is an ingredient of jerk pastes, used for marinating meat.

Nº 21

Origanum vulgare subsp. *hirtum*

oregano
by Mary Maitland (Mrs George Govan),
Kew Collection, c.1823–32

Several species have an oregano-like flavour, aromatic and warm. This subspecies is native to Greece and Turkey and the leaves are most used in Mediterranean foods such as pizza.

Nº 22

Cinnamomum verum

cinnamon, Ceylon cinnamon
by unknown Indian artist commissioned by
William Roxburgh, Kew Collection, c. 1790

The dried inner bark of several related species can be used for cinnamon; this species is sweeter, while cassia bark (*C. cassia*) has a bitter edge. Sri Lanka is home to the wild forms of *C. verum*, and the largest exporter of the cultivated form. Used whole (as quills) or powdered, the spice has an aromatic, sweet flavour. It is widely used in South Asian curries, and in baked goods and confectionary in Europe.

Nº 23

Salvia rosmarinus

rosemary
by Elizabeth Blackwell from
Elizabeth Blackwell *A Curious Herbal*, 1737–9

Rosemary leaves are a popular herb in the Mediterranean region, both fresh and dried. Rosemary goes well with roast meat and vegetables, but its aromatic flavour can be overwhelming if used to excess.

Nº 24

Cuminum cyminum

cumin

from F. G. Hayne *Getreue Darstellung und Beschreibung der in der Arzneykunde Gebräuchlichen Gewächse*, 1805–46

Cumin was probably first taken into cultivation in the Near East in prehistoric times. The fruits of this small annual plant are strongly aromatic and improve even further on toasting. They are used worldwide, particularly in Indian cuisine, and in spice mixes in South America.

N° 25

Murraya koenigii

curry leaf

by unknown Indian artist commissioned by William Roxburgh, Kew Collection c.1800

Curry leaves grow on a small tree native to warmer parts of Asia. Always used fresh, they have a strong, aromatic flavour and are a vital component of curry dishes in southern India and Sri Lanka. They are briefly fried before use.

Nº 26

Syzygium aromaticum

clove

from the Marianne North Collection, Kew, 1883

Cloves are the flower buds of an evergreen tree native to the Malaku (Moluccan) Islands , an archipelago in Indonesia. The essential oil eugenol, also found in cinnamon and allspice, gives cloves a powerfully aromatic and warming taste. Cloves have been used in Chinese cooking for at least 2,500 years; uses in Europe include as a pickling spice and an ingredient in festive foods such as Christmas pudding.

Nº 27

Trigonella foenum-graecum

fenugreek

by Ferdinand Bauer from John Sibthorp
Flora Graeca, 1806–40

Fenugreek is best known for its small brownish-yellow seeds, but the leaves are used as greens in India and other countries. It has been cultivated in its area of origin, the eastern Mediterranean and Near East, for at least 5,000 years. The bitter, aromatic seeds are fried in oil and added to Indian curries. Fenugreek and coriander seeds dating back to approximately 3,300 years ago were found in the tomb of Tutankhamun.

Nº 28

Zingiber officinale

ginger

by unknown Indian artist commissioned by William Roxburgh, Kew Collection c.1800

Ginger is probably native to India, but is now widely cultivated throughout the wet tropics. The rhizome (root) has a pungent, warm flavour and is used fresh in Asian cooking. It has a long history of export in dried form, reaching Europe in Roman times. In European cuisine powdered ginger has been used in baked goods such as gingerbread since medieval times.

Nº 29

Juniperus communis

juniper

by Mary Anne Stebbing, Kew Collection, 1897

The berry-like cones of juniper gain their aromatic, sweet scent from essential oils including pinene and thujone. The cones take three years to mature and are gathered by hand from wild trees in Europe, the main region in which juniper is harvested and used. Its culinary uses include meat dishes, and as the defining component of gin, as well as a low-alcohol beer made in northern Poland.

Nº 30

Vanilla planifolia

vanilla

from C. A. Lemaire

L'horticulteur Universel 1839–45

Vanilla is the ripe fruit of a tree-climbing vine in the orchid family, native to Central America and northern South America. After harvest the pods are blanched, fermented and dried, during which time the sweet, aromatic scent develops. The Comoros Islands, Madagascar and Indonesia are now major exporters, but when cultivated outside its native home, vanilla must be pollinated by humans, using a small stick.

Nº 31

Citrus hystrix

makrut lime

from H. L. Duhamel du Monceau
Traité des Arbres et Arbustes que l'on Cultive en France en Pleine Terre, 1800–19

The leaves of makrut lime have a winged stalk of a similar size and shape as the leaf, leading to an unusual figure-of-eight shape. The leaves have a strong lemon aroma and are much used in soups and stir-fries in Thailand and nearby countries. The fruits are inedible, but the rind can be grated for culinary use.

Nº 32

Crocus sativus

saffron

possibly by Vishnupersaud, commissioned by John Forbes Royle, Kew Collection, c.1828

The orange-red styles and stigmas (female parts) of the crocus flower are picked by hand, a labour-intensive process making it the most expensive of all spices. Saffron was first cultivated in Crete, where frescoes at Knossos dating to 1700 BCE show monkeys making the harvest. Saffron has a pungent, spicy taste and imparts a deep yellow colour. It is only used in minute quantities.

Nº 33

Myristica fragrans

nutmeg and mace

from Charles Morren *Belgique Horticole*, 1856

Two spices come from the fruit of the nutmeg tree: nutmeg, which is the hard, brown seed, and mace, the leathery red aril that wraps around the seed. Like cloves, nutmeg trees are native to the Malaku (Moluccan) Islands of Indonesia, which are still the largest exporter. Both spices are aromatic and warm in flavour. Nutmeg should be grated immediately before use as the powder rapidly loses its aroma.

Nº 34

Sesamum indicum

sesame
by unknown Indian artist commissioned by William Roxburgh, Kew Collection c.1800

The oil-rich seeds of sesame have a nutty flavour, intensified by toasting. Sesame is an annual crop that was first grown in India, about 5,000 years ago. Cultivation has now spread to warm regions worldwide. Notable food uses include halva and tahini in the eastern Mediterranean, as a cooking oil in India and China, and in diverse uses of the white or black seeds in Japanese cuisine.

Nº 35

Allium schoenoprasum

chives

from James Sowerby *English Botany*, 1869

Wild chives have a long history of use in Europe and seem to have entered cultivation some 500 years ago. They are always eaten fresh and have a delicate onion flavour. The leaves are used in sauces and as a garnish. Related species have similar uses in China.

Nº 36

Piper nigrum

pepper

by unknown Indian artist commissioned by William Roxburgh, Kew Collection c.1800

Peppers are the fruits of a woody climber, native to the Ghat mountains of southern India, but with a long history of wide cultivation. Peppers were the most important product of the spice trade, reaching Europe in Roman times, and much used in food since medieval times. Green peppercorns are unripe, black are unripe and whole, and white are picked when ripe and have had the skin of the fruit removed.

Nº 37

Armoracia rusticana

horseradish

from Jan Kops *Flora Batava*, 1822

The roots of horseradish contain chemicals related to those in mustard, giving a similar strongly pungent odour and taste. It was probably first cultivated in medieval times, perhaps in the eastern Mediterranean. In Europe it is used grated or as a relish with savoury dishes.

Nº 38

Glycyrrhiza glabra

liquorice
from François Pierre Chaumeton
Flore Médicale Décrite, 1815–20

The liquorice plant grows wild in the eastern Mediterranean and has been cultivated in Europe for about 500 years. The roots contain glycyrrhizin, which is 50 times sweeter than sugar. 4,000 years ago liquorice was used as a medicine and for desserts in ancient Mesopotamia; today it is used worldwide as a flavouring in sweet foods, and medicinally including in cough mixtures. Over-consumption of liquorice can raise blood pressure to dangerous levels.

Nº 39

Perilla frutescens

shiso, beefsteak plant
by J. Curtis from
Curtis's Botanical Magazine, 1823

Shiso is a member of the mint family, with aromatic leaves that are used as a garnish in South-East Asian cuisine. In Japan red-leaved forms are used to add colour to pickles such as pickled ginger.

N° 40

Tamarindus indica

tamarind

probably by Rungiah commissioned by Robert Wight, Kew Collection, c.1825–28

Tamarind trees are native to tropical Africa and are now grown throughout the tropics for their pods. These contain a dark pulp that is sour and fruity. It is much used in South Indian cuisine, in South-East Asia, and as the basis of a refreshing drink in the Caribbean and South America.

ILLUSTRATION SOURCES

Books and Journals

Blackwell, E. (1737–9). *A Curious Herbal, containing five hundred cuts, of the most useful plants, which are now used in the practice of physick.* 2 volumes. J. Nourse, London.

Blackwell, E. (1750–73). *Herbarium Blackwellianum.* 6 volumes. Typis Io. Iosephi Freischmanni, Nuremberg.

Bulliard, P. (1776–1783). *Flora Parisiensis, ou, Descriptions et Figures des Plantes qui Croissent aux Environs de Paris.* 6 volumes. P. Fr. Didot le Jeune, Libraire, quai des Augustins, Paris.

Chaumeton, F. P. (1815–20). *Flore Médicale Décrite.* 7 volumes, Panckoucke, Paris.

Descourtilz, M. É. (1821–9). *Flore Pittoresque et Médicale des Antilles.* 8 volumes. Chez Corsnier, Paris.

Curtis, W. *et al.* (1798). *Curtis's Botanical Magazine.* Volume 12, t. 400.

Duhamel du Monceau, H. L. (1800–19). *Traité des Arbres et Arbustes que l'on Cultive en France en Pleine Terre.* 7 volumes. Didot *et. al.*, Paris.

Hayne, F. G. (1805–46). *Getreue Darstellung und Beschreibung der in der Arzneykunde Gebräuchlichen Gewächse,* Auf Kosten des Verfassers, Berlin.

Köhler, F. E (1887). *Medizinal-Pflanzen: in naturgetreuen Abbildungen mit kurz erläuterndem.* 2 volumes. F.E. Köhler, Gera-Untermhaus.

Kops, J. (1800–46). *Flora Batava.* J. C. Sepp en Zoon, Amsterdam.

Lemaire, C. L. (1839–45). *L'horticulteur universel : journal général des amateurs et jardiniers présentant l'analyse raisonée des travaux horticoles français et étrangers.* 7 volumes. H. Cousin, Paris.

L'Obel, M. de, Dodoens, R., Clusius, C. (1581). *Plantarum seu Stirpium Icones.* 2 volumes. C. Plantini, Antwerp.

Morren, C. (1851–85). *La Belgique Horticole : Annales de Botanique et d'Horticulture.* La Direction Générale, Liége.

Plenck, J. J. R. von. (1788–1812). *Icones Plantarum Medicinalium.* 8 volumes. Vienna.

Prain, D. (1918). *Curtis's Botanical Magazine.* Volume 144, t. 8754.

Rumphius, G. E. (1750). *Herbarium Amboinense ... in Latinum sermonem Versa Cura et Studio J. Burmanni.* 6 volumes. Apud Meinardum Uytwerf, Amsterdam.

Sibthorp, J. (1806–40). *Flora Graeca.* 12 volumes. Richardi Taylor et socii, London.

Sims, J. (1823). *Curtis's Botanical Magazine.* Volume 50, t. 2395.

Sowerby, J. E. (1863–1886). *English Botany, or, Coloured Figures of British Plants.* 12 volumes. Hardwicke, London.

Spach, Édouard. (1834–48). *Histoire Naturelle des Végétaux.* 14 volumes. Librairie Encyclopédique de Roret, Paris.

Art Collections

Indian Botanical Art Collections – traditionally known as Company School (or *Kampani kalam*), commissioned by employees of the East India Company from largely unknown Indian artists in the 18th and 19th centuries, showing considerable stylistic diversity and beauty. Most were given to Kew in 1879 when the museum and library of the East India Company was dispersed, following its inheritance by the India Office of the British government. Collections represented in this book:

Adam Freer – almost 400 drawings made in Bengal, the earliest dated 1793, with over half unsigned, but some 150 ascribed to seven artists; Muslims Mogul-Ian and Mirza-Ian and five Hindu all bearing the name Lall.

Mary Maitland (Mrs George Govern) – drawings of mainly Indian plants dated 1823–32, by Indian artists, also some by Mary Maitland, whose husband, George Govan, was the First Superintendent of the Botanic Garden in Saharunpur (1819–23). The collection was donated to Kew in 1904 by her sons.

William Roxburgh 'Icones' – one of two sets of comprising 2,500 drawings by Indian artists, made 1776–1813, working on the Coromandel Coast and at the Calcutta Botanic Garden. A duplicate set is held by the Central National Herbarium, Botanical Survey of India, in Kolkata's Acharya Jagadish Chandra Bose Indian Botanic Garden.

John Forbes Royle – noted for his *Illustrations of the Botany of the Himalayan Mountains* (1833–1840). The great botanical artist Vishnupersaud made illustrations for Royle, arguably the most beautiful of all the illustrations of Indian plants ever made.

Robert Wight – detailed drawings of plants, mainly from southern India, by Indian artists Rungiah and his pupil Govindoo for Wight's published works including *Spicilegium Neilgherrense* (1846–51) and *Icones Plantarum Indiae Orientalis* (1838–1853).

Marianne North (1830–90) – comprising over 800 oils on paper, showing plants in their natural settings, painted by North, who recorded the world's flora during travels from 1871 to 1885, with visits to 16 countries in 5 continents. The main collection is on display in the Marianne North Gallery at Kew Gardens, bequeathed by North and built according to her instructions, first opened in 1882.

Mary Anne Stebbing (1845–1927) – botanist and illustrator, daughter of botanist and entomologist William Wilson Saunders and wife of zoologist Thomas Roscoe Rede Stebbing. One of the first women to be admitted to the Linnean Society of London. A set of her illustrations was presented to Kew by E. C. Wallace Esq., December, 1946.

FURTHER READING

Bown, D. (2002). The *Royal Horticultural Society New Encyclopedia of Herbs & Their Uses*. Dorling Kindersley, London

Dalby, A. (2000). *Dangerous Tastes: The Story of Spices*. British Museum Press, London and University of California Press, Berkeley.

Dauncey, E. A. and Howes, M.-J. (2020). *Plants that Cure: Plants as a Source of Medicines, from Pharmaceuticals to Herbal Remedies*. Royal Botanic Gardens, Kew.

Davidson, A. (2014). *Oxford Companion to Food, third edition*. Oxford University Press, Oxford.

Farrimond, S. (2018). *The Science of Spice: Understand Flavour Connections and Revolutionize your Cooking*. Dorling Kindersley, London.

Freedman, P. (2008). *Out of the East: Spices and the Medieval Imagination*. Yale University Press, New Haven.

de Guzman, C.C. and J.S. Siemonsma (eds). (1999). *Plant Resources of South East Asia No. 13. Spices.* Backhuys, Leiden.

Norman, J. (2015). *Herbs and Spices: The Cook's Reference*. Dorling Kindersley, London

North, Marianne and Mills, Christopher. (2018). *Marianne North: The Kew Collection.* Royal Botanic Gardens, Kew.

Prance, G., and Nesbitt, M. (eds). (2005). *The Cultural History of Plants*. Routledge, New York.

Payne, Michelle. (2016). *Marianne North: A Very Intrepid Painter*. Revised edition. Royal Botanic Gardens, Kew.

Rix, M. (2021). *Indian Botanical Art: An Illustrated History.* Royal Botanic Gardens, Kew and Roli Books, New Delhi.

Rumphius, G. E. and Beekman, E. M. (2011). *The Ambonese Herbal*. 6 volumes. National Tropical Botanical Garden and Yale University Press, New Haven and London.

Simmonds, M., Howes, M.-J., Irving, J. (2016). *The Gardeners Companion to Medicinal Plants.* Frances Lincoln, London in association with the Royal Botanic Gardens, Kew.

Turner, J. (2004). Spice: *The History of a Temptation*. HarperCollins, London.

Van Wyk, B.-E. (2013). *Culinary Herbs and Spices of the World*. Royal Botanic Gardens, Kew and University of Chicago Press, Chicago.

Willis, Kathy and Fry, Carolyn. (2014). *Plants from Roots to Riches.* John Murray, London in association with the Royal Botanic Gardens, Kew.

Online

http://gernot-katzers-spice-pages.com – Gernot Katzer's spice pages, an indispensable resource.

www.biodiversitylibrary.org – the world's largest open access digital library specialising in biodiversity and natural history literature and archives, including many rare books.

www.kew.org – Royal Botanic Gardens, Kew website with information on Kew's science, collections and visitor programme.

www.plantsoftheworldonline.org – an online database providing authoritative information of the world's flora gathered from the botanical literature published over the last 250 years.

ACKNOWLEDGEMENTS

Kew Publishing would like to thank the following for their help with this publication: Elizabeth Dauncy, Henry Noltie, Kim Walker; in Kew's Library and Archives, Fiona Ainsworth, Craig Brough, Julia Buckley, Anne Marshall, Lynn Parker; for digitisation work, Paul Little.

INDEX

Royal Botanic Gardens Kew

First published in 2022
Royal Botanic Gardens, Kew,
Richmond, Surrey, TW9 3AB, UK
www.kew.org

ISBN 978 1 84246 753 4

Distributed on behalf of the Royal Botanic Gardens, Kew in North America by the University of Chicago Press, 1427 East 60th St, Chicago, IL 60637, USA.

British Library Cataloguing in Publication Data
A catalogue record for this book is available from the British Library

Design: Ocky Murray
Page layout: Christine Beard
Image work: Nicola Erdpresser and Christine Beard
Production Manager: Jo Pillai
Copy-editing: Michelle Payne

Printed and bound in Italy by Printer Trento srl.

Front cover images: *Tamarindus indica*, tamarind, *Cinnamomum verum*, cinnamon from the Roxburgh Collection, Kew; *Piper nigrum*, pepper, *Capsicum annuum*, chilli, *Mentha crispa*, mint from Plenck *Icones Plantarum Medicinalium*; *Thymus vulgaris*, thyme from Duhamel du Monceau *Traité des Arbres et Arbustes*; *Petroselinum crispum*, parsley from Köhler *Medizinal-Pflanzen*; *Illicium verum*, star anise from Chaumeton *Flore Médicale.*

Endpapers: *Rhus coriaria* sumac and *Salvia rosmarinus* rosemary from H. L. Duhamel du Monceau *Traité des Arbres et Arbustes que l'on Cultive en France en Pleine Terre*, 1800–19.

p.2: *Citrus hystrix* makrut lime from Georgius Everhardus Rumphius *Herbarium Amboinense*, 1750.

p.4: *Zingiber officinale* ginger from Édouard Spach *Histoire Naturelle des Végétaux*, 1834–48.

p.10–11: Old tamarind trees, India. Sketch by Joseph Dalton Hooker, Kew Collection, 1848.

For information or to purchase all Kew titles please visit shop.kew.org/kewbooksonline or email publishing@kew.org

Kew's mission is to understand and protect plants and fungi, for the wellbeing of people and the future of all life on Earth.

Kew receives approximately one third of its funding from Government through the Department for Environment, Food and Rural Affairs (Defra). All other funding needed to support Kew's vital work comes from members, foundations, donors and commercial activities, including book sales.

Publishers note about names

The scientific names of the plants featured in this book are current, Kew accepted names at the time of going to press. They may differ from those used in original-source publications. The common names given are those most often used in the English language, or sometimes vernacular names used for the plants in their native countries.